# MYSTERY "OLD SMOKER" STARS FOUND IN THE MILKY WAY:

## Galactic Search For Protostars Yields An Unexpected Stellar Discovery - Unraveling The Puzzle

**MELISSA P. THURMAN**

# TABLE OF CONTENTS:

# INTRODUCTION

For decades, we've seen big red stars as lovely sunset painters, fading into obscurity. They inflate, cool, and shed their outer layers in wispy breezes, replenishing the interstellar medium with life-giving ingredients.

However, the "old smokers" are cosmic anarchists, which disrupts this neat narrative. Their outbursts are fierce, unexpected, and cloaked in mystery. What drives these outbursts? What secrets do they keep inside their dusty hearts?

These questions spark a range of ideas. The "old smokers" might be nature's cosmic recyclers, with their puffed-out dust clouds containing heavy elements required for the development of subsequent stars and planets. These enriched clouds might be the cradles of new planets, with the possibility for life to bloom in the shadows of these spectacular stars. However, their effect goes beyond star and planet formation.

The "old smokers" are cosmic storytellers, with stories of galaxy development engraved in their dust.

By examining these whispers, we may be able to recreate the history of our Milky Way, comprehending the dance of star birth and death, and tracing the genealogy of the components that comprise our planet, our bodies, and the oxygen we breathe.

The quest for these cosmic curiosities goes beyond our galaxy. The search for "old smokers" in other, different galaxies may reveal a world of cosmic variety beyond anything we've ever seen. Stars with distinct compositions, histories, and celestial dances may have their versions of "old smokers," telling unique tales about their cosmic history and the possibility of life in their areas.

The "old smokers" are more than simply cosmic curiosities; they represent a link between the known and the unknown, a portal to a world much more vibrant than we ever dreamed.

They inspire awe, a desire to explore, and a newfound respect for the interconnection of everything. Their echoes reveal a cosmic truth: the cosmos is full of surprises, and its mysteries await those who dare to listen, question, and start on the wonderful road of discovery.

# CHAPTER 1: INTRODUCING THE VISTA SURVEY AND ITS AMBITIOUS GOALS.

The night sky, a world of glittering splendor, has enthralled humans since the beginning of time. Its mysteries enticed us, whispering stories of heavenly bodies and secrets about our existence. However, our vision of this cosmic extravaganza has always been constrained by the curtain of visible light.

Under the surface, many dramas unfold, star births and deaths, cosmic dances of dust and gas, all hidden from our naked sight. With this desire to break through the cosmic curtain, scientists began on a momentous journey known as the Visible and Infrared Survey Telescope for

Astronomy (VISTA). The Cerro Paranal Observatory in Chile houses VISTA, a technical wonder. Its 4.1-meter mirror, made of lightweight honeycomb tiles, collects light with incredible efficiency.

VISTA, unlike human eyes, sees beyond the visible light spectrum and into the infrared. This invisible radiation, released by heated dust and cold stars, opens up a secret world, a huge area teeming with celestial objects unseen to the naked sight.

The VISTA survey started on an unprecedented mission: to map roughly a billion stars in the center of the Milky Way galaxy. This was an extraordinary endeavor, a cosmic census that dwarfed any prior attempt.

Consider wandering a crowded urban metropolis with a billion people, endeavoring to observe their lives, movements, and relationships. The sheer numbers would be overwhelming, yet VISTA, like a diligent astronomer, proceeded on its celestial metropolis study.

The objectives of VISTA were as lofty as its breadth. One of the primary goals was to discover the mysteries of nascent stars, which are shrouded under thick clouds of gas and dust. Visible light struggles to penetrate these galactic nurseries, yet infrared photons glide readily through the cosmic dust, exposing protostars' secret brightness.

By monitoring the brightness fluctuations of thousands of stars, VISTA planned to catch these stellar children in the throes of growing, watching their outbursts and eruptions, and studying their dance with surrounding gas as they build their planetary systems.

Another ambitious objective was to investigate the galaxy's dying embers, the red giants, which are stars reaching the end of their nuclear furnaces. These massive, bloated, crimson entities spew material into space, replenishing the interstellar medium with the ashes of their nuclear furnaces.

VISTA, with its capacity to peek through dust clouds, hoped to observe these dying breaths, chart the patterns of expelled material, and better grasp the role of red giants in the life and death of the cosmic cycle. But the ultimate objective of VISTA extended beyond particular stars. It aimed to provide a comprehensive snapshot of our galaxy, mapping the rich fabric of dust clouds, gas filaments, and star populations.

VISTA attempted to unravel the Milky Way's chemical history by documenting the distribution of elements inside it, dating back to the cataclysmic deaths of its earliest stars. This map envisioned as a cosmic Rosetta Stone, would provide the key to understanding the Milky Way's genesis and development, as well as the possibility of life-bearing planets inside its spiral arms.

The VISTA survey was more than just a technical marvel; it was a monument to human curiosity, a joint endeavor by astronomers from throughout the world.

From Europe's busiest observatories to the isolated outpost of Cerro Paranal, astronomers worked together to gain cosmic knowledge. They worked through mounds of data, analyzing the light whispers caught by VISTA, and piecing together the complicated narrative of our cosmic neighborhood.

As data flowed in, VISTA's view showed a cosmos that was considerably richer and more active than previously thought. Newborn stars burst spectacularly, red giants emitted plumes of dust, and hitherto unknown patterns appeared from the cosmic turmoil. The Milky Way, which was previously a static background, came to life, pulsing with energy and full of surprises.

This is just the beginning of VISTA's legacy. The material it acquired, a wealth of cosmic knowledge, awaits further investigation. As astronomers probe further into this cosmic record, they will undoubtedly meet unexpected events, uncover new secrets, and rewrite our view of the cosmos.

VISTA's billion-star survey is more than a map; it is an invitation, a beacon calling us to continue our cosmic journey, to push the limits of knowledge, and to enjoy the unfathomable beauty that lies within the sparkling world of the night sky.

# CHAPTER 2: TRACKING ELUSIVE NEWBORN STARS AND ENCOUNTERING THE UNEXPECTED.

Under the velvety shroud of darkness, Newborn stars, called Protostars, cloaked in cosmic nurseries of dust and gas, spring into creation, their youthful enthusiasm flashing brief glimmers against the background of the Milky Way. These protostars, which are just a million years old, are the universe's newest recruits, the creators of future planetary systems.

But their birth pains are often shrouded in mystery, obscured by the cosmic dust that surrounds them. This is when the VISTA survey and our tale take a dramatic turn, delving into the domain of the invisible, the search for these hidden giants. VISTA, with its acute infrared vision, became the cosmic bloodhound, sniffing out the celestial babes' secret brightness. Unlike visible light, infrared photons glide through interstellar dust easily, showing the unmistakable marks of star adolescence such as energy outbursts, dramatic brightness surges, and the shaping of surrounding gas as planets form.

Imagine a detective narrative set in the vastness of the Milky Way. Our characters are astronomers, equipped with a large telescope and a sharp eye for cosmic hints. The suspects are concealed protostars in dusty cocoons.

What's the evidence? Flickers of infrared light, abrupt shifts in brightness, and the unmistakable signs of young planets revolving around their mysterious mothers.

The hunt started with rigorous preparation. The VISTA team, like great cartographers, methodically split the galactic center into regions, with each pixel potentially harboring a celestial newborn.

Night after night, the telescope meticulously scanned these celestial quadrants, obtaining pictures of extraordinary clarity. Every brightness shift and infrared flicker was methodically documented, resulting in a cosmic census of possible protostars. However, identifying the genuine offenders among the heavenly din was difficult.

Other astronomical phenomena, such as pulsing red giants and variable stars, resembled protostar outbursts. Disentangling the genuine identities required rigorous investigative work. Spectral research, similar to fingerprinting the light, showed the distinct characteristics of young stars - the telltale ratios of components formed in their burning cores. Through this spectral detective work, the astronomers meticulously found 32 of these hidden giants, catching their eruptions in real-time, a first in cosmic history.

Observing these cosmic outbursts was a show in itself. Some protostars brightened by a factor of 40, while others increased by a whopping 300 times.

Their eruptions, fuelled by the accumulation of surrounding gas, lasted months, years, or even decades. Astronomers studied these light curves to record the development spurts of these celestial children, witnessing their dance with surrounding gas as they shaped the whirling disks that would one day form planetary systems.

However, VISTA's glance revealed more than simply what was anticipated. Among the 32 verified protostars, 21 intriguing objects resisted straightforward classification. They showed odd fading patterns, unlike the characteristic eruptions of infants. Were they hiding giants in disguise? Or something completely different?

This unexpected meeting began a new chapter in the cosmic detective narrative, sending the astronomers on a road of mystery and adventure.

These 21 celestial anomalies, situated in the galactic core, stayed stubbornly dim for years before unexpectedly plunging into even deeper darkness for months before gradually resurfacing. Could these be protostars in a particularly bashful state? Or were they red giants, the galaxy's old stars, entering a cosmic twilight?

Further examination, similar to a witness questioning in the heavenly courts, revealed their genuine identities. Spectral research revealed that these were not protostars, but rather an unknown form of red giant.

These "old smokers," as they were appropriately termed, were silent for years until suddenly exhaling massive clouds of dust and gas, briefly hiding their brightness. This smoking behavior contradicted the traditional understanding of red giants, putting a kink in existing ideas of star death.

The finding of the aged smokers demonstrated VISTA's capacity to reveal the unexpected. It served as a reminder that the cosmos is full of surprises, with mysteries buried in plain sight, ready to be revealed with the correct tool and perspective. These "old smokers" were more than just celestial curiosities; they were potential game changers, raising issues about the distribution of elements in the galaxy, the causes of star death, and perhaps the beginnings of life itself.

The quest for hidden giants was more than merely checking boxes on a cosmic checklist. It was a quest of discovery, demonstrating the human spirit's infinite curiosity. VISTA, like a cosmic bloodhound, led the way, revealing the secrets of young and ancient stars and rewriting our knowledge of the cosmos with each flicker of infrared light it collected.

.

# CHAPTER 3: DISCOVERING THE "OLD SMOKERS" AND THEIR PUZZLING BEHAVIOR

Deep inside the Milky Way, among a whirling world of stars and dust, the VISTA telescope discovered a cosmic anomaly, a celestial enigma that defied expectations and prompted a frenzy of scientific curiosity. The brilliant protostars that lit up the intergalactic nursery weren't your ordinary noisy children. These were the "old smokers," enigmatic crimson giants shrouded in darkness and known for their mysterious puffs of smoke.

Imagine this scene: A team of astronomers is combing over data from VISTA, a massive database of cosmic observations.

Day after day, they methodically examine the light curves of stars, tracking their brightness, and seeking patterns, and clues to the universe's great narrative. The 32 verified protostars, with their spectacular outbursts, were wonderful discoveries, but something unexpected stood out: 21 stars remained stubbornly dim, cloaked in mystery.

Years, months, and days passed on the cosmic clock, yet these 21 heavenly mysteries remained unsolved. Then, like a magician pulling a rabbit out of a hat, something unexpected occurred. These peaceful giants vanished in a spectacular blast of smoke, fading into near-oblivion for months before gradually reverting to their old dull condition.

It was as if they had gone, enveloped in a weird cosmic fog. This odd behavior caused waves across the astronomy community. The traditional ideas of red giants, the universe's dying embers, simply could not explain such spectacular smoke signals.

In their final years, red giants were meant to decay and spew material at a more predictable, almost leisurely pace. These "old smokers," on the other hand, were upsetting the balance, testing our knowledge of star death and whispering secrets about the galaxy's inner workings.

To solve the puzzle, the astronomers needed to look closer. Spectral analysis, similar to deciphering the fingerprints of light, became their primary instrument.

By examining the composition of the released gas, scientists identified a distinct signature - a combination of heavy components considerably richer than predicted from normal red giants. This abundance hinted at a distinct genesis story, a new chapter in the cosmic narrative.

These "old smokers," it turns out, were most likely created in the galactic center's metal-rich furnace. This one-of-a-kind atmosphere, rich with materials formed by previous generations of stars, gave them an additional boost, resulting in their unusual behavior. The heavier elements, functioning as cosmic dust bunnies, might cluster together more readily, producing thick disks surrounding the dying stars.

These disks, in turn, maybe the source of the mysterious smoke signals, sometimes concealing the star as they wobbled or were expelled in spectacular outbursts.

But the questions continued to pile up. What prompted the smoke signals? Was there a gravitational tug-of-war with a secret binary companion? Or a dramatic blip in the star's inner furnace? The solutions were cloaked in cosmic fog and remained elusive.

The finding of the "old smokers" was not just an astronomical curiosity; it was a game changer. These mysterious stars give insight into the distribution of heavy metals in the galaxy, which may influence the development of future stars and planets.

Their narrative is still being written, with each fresh observation adding another layer to the enigma.

As telescopes get more powerful and scientists go further into the galactic core, the mysteries of the "old smokers" may ultimately be revealed. Their smoke signals, previously a cosmic mystery, might become lighthouses pointing us to a better comprehension of the universe's inner workings, reminding us that hidden giants hide in the darkness, waiting to be uncovered and rewrite the tales we thought we knew.

# CHAPTER 4: THEORIES AND SPECULATIONS ON THE MECHANISM BEHIND THE SMOKE CLOUDS.

The "old smokers" of the Milky Way, identified by the VISTA survey, have become a cosmic mystery, with their unexplained puffs of smoke contradicting traditional knowledge of red giants. These old stars, cloaked in the galactic center's deep dust lanes, have sparked a cosmic detective thriller, with scientists eager to pursue ideas and guesses to explain their baffling behavior.

One crucial issue hangs heavy in the intergalactic air: what exactly causes these catastrophic dust and gas eruptions?

Is it just a quirk of their nature, a distinct trait created in the furnace of the galactic core, or a sign of some underlying stellar drama? While definite solutions remain difficult, a plethora of intriguing theories have surfaced, each throwing light on a different part of the mystery.

**The Dusty Disk Theory:** This concept posits that the "old smokers" include thick disks of circumstellar debris that are rich in heavy elements owing to their galactic center origins. These disks, unlike those orbiting younger stars, may be more prone to instabilities because of their unequal composition. Consider a celestial dancer with one leg made of lead and the other of feathers; each spin may cause a wobble, allowing dust particles to separate and obscure the star in brief smoke clouds.

**The Binary Dance:** Could "old smokers" be part of a cosmic dance, engaged in a gravitational tango with unknown binary companions? The gravitational tug-of-war between these celestial partners may sometimes rupture the star's outer layers, resulting in the outflow of dust and gas in the form of magnificent smoke signals. This idea might account for the irregular character of the outbursts since the orbital dance determines the frequency and severity of gravitational tugging.

**Stellar Hiccups:** Perhaps the answer is found within the stars themselves. Red giants in their latter years have unpredictable internal processes. Convective currents swirl across their massive stomachs, sometimes generating glitches in the nuclear furnace. These abrupt bursts of activity may expel material from the star's outer layers, resulting in brief smoke clouds. This explanation is consistent with the observed darkening of "old smokers" during puffing episodes when the expelled material blocks out their light.

**Hidden Companions:** Another fascinating theory is the existence of undetected gas giants lurking on the outskirts of "old smoker" systems. These celestial behemoths, unseen to modern observatories, might stir up the circumstellar disk with their gravitational impact, causing dust ejections and leading to irregular light curves. This notion, although fascinating, is still very speculative, awaiting observable confirmation of these hypothetical hidden giants.

**Magnetic mysteries:** Some astronomers believe that the magnetic field of "old smokers" may play a significant role in the dusty drama. Red giants have intricate magnetic fields that may direct and manipulate charged particles inside their outer layers. Sudden variations in the magnetic field configuration may operate as a cosmic wind, blowing away dust and gas and producing observable smoke clouds. This proposal, however, needs further theoretical and empirical research to properly comprehend the interaction between magnetic fields and dust outflow in aging stars.

The effort to solve the mystery of the "old smokers" involves more than simply fulfilling our cosmic curiosity. It has the potential to fundamentally alter our knowledge of star development, especially in the metal-rich environment of the galactic core. Their smoke signals might serve as a guidepost, leading to new hypotheses on dust formation, planetary system dynamics, and the development of red giants in severe settings.

As with any good detective novel, fresh clues are continuously revealed. Further studies with sensitive telescopes at infrared and other wavelengths may disclose the telltale signs of hidden partners, establish the existence of dusty disks, or expose the inner workings of the stars by spectroscopic research. Each fresh piece of evidence, each cosmic whisper caught, moves us closer to breaking the code of the "old smokers," rewriting the chapter on red giant behavior, and revealing the mysteries they contain about the Milky Way's history and future.

## CHAPTER 5: EXPLORING THE ROLE OF HEAVY ELEMENTS IN THE GALACTIC CENTER AND THEIR INFLUENCE ON "OLD SMOKERS".

Deep inside the churning core of the Milky Way, in a cosmic dance of dust and light, lies the tale of the "old smokers" - mysterious red giants whose strange puffs of smoke call into question our knowledge of star death. This chapter, however, goes beyond the smoke into the invisible ballet - the dance of heavy elements in the galactic core and their tremendous impact on these cosmic anomalies.

Imagine entering a cosmic ballroom made of whirling dust and throbbing stars, rather than silk and crystal.

Gravity serves as the orchestra conductor, controlling the motions of billions of heavenly bodies, while the dancers' elemental makeup determines the complex steps of their cosmic waltz. At the heart of this dance, bathed in the strong brilliance of young stars and shrouded in the relics of their predecessors, lay the "old smokers," their movements controlled by the particular concoction of heavy elements they acquired from their fiery origins.

These heavy elements, formed in the hearts of previous generations of supergiants, serve as the heavenly ballerinas' weighted skirts. Denser elements, such as iron, silicon, and calcium, cluster together more easily, producing thick disks surrounding the "old smokers".

These disks, unlike those orbiting younger stars, are more sensitive to gravitational tugging and internal instabilities, which might result in the violent ejections of dust and gas that characterize "old smoker" puffs.

Consider a ballerina with one leg made of lead and the other from feathers. Every spin, every change in the music causes an imbalance, a possible stumble. Similarly, the uneven distribution of heavy elements in the "old smoker" disks may cause them to wobble, resulting in dust pieces detaching and obscuring the star in transient smoke clouds.

This "dusty disk theory" proposes that the galactic center's richness, or quantity of heavy elements, is the key to explaining the "old smokers" mysterious behavior. The dance of weighty pieces is more than simply clumping and wobbling, however. It's also a narrative about gravitational pulls and cosmic tangos. According to some scientists, the "old smokers" may be dancing with unseen binary counterparts in space.

These secret companions, hiding in the darkness, might exert gravitational force on the star and its surrounding disk, causing periodic interruptions that result in smoke signals. The frequency and severity of these outbursts would therefore be determined by the binary system's orbital dance, adding another level of intricacy to the cosmic ballet. However, the dance of heavy elements affects more than just the surface layers of "old smokers"; it may also have an impact on their interior functioning.

Consider a cosmic hiccup—a sudden rise in the star's nuclear furnace. These powerful belches, induced by instabilities in the dying star's heart, may release material from its outer layers, resulting in brief smoke clouds. This "stellar hiccups theory" proposes that heavy metals, by changing the interior processes of stars, may lead to their unpredictable behavior and spectacular smoke signals.

The heavy components also have a significant influence on the role of "old smokers" in the Milky Way's grand tale.

These mysterious stars, formed in the galactic center's metal-rich furnace, serve as cosmic couriers, conveying their enhanced composition further out into the spiral arms. As they shed their dusty robes, they disperse heavy elements across the interstellar medium, nourishing the cosmic soup from which future generations of stars and planets will emerge.

This "elemental legacy" argument suggests that "old smokers" may have a greater role. Their smoke signals, although seeming to be a celestial anomaly, might constitute an important phase in the galactic recycling cycle, guaranteeing the continuing enrichment of the Milky Way and perhaps affecting the circumstances for life's formation on future planetary systems.

Unraveling the dance of heavy elements inside the "old smokers" is more than simply solving a stellar puzzle; it's about comprehending the complex interaction between composition, environment, and astronomical dynamics. Every observed puff of smoke, every shift in the light curve, has the potential to rewrite our understanding of red giants, shed light on the unique conditions of the galactic center, and reveal the hidden choreography that shapes galaxies' evolution and the possibility of life within them.

# CHAPTER 6: THE CONTRIBUTION OF "OLD SMOKERS" TO THE INTERSTELLAR DUST AND ITS IMPACT ON STAR FORMATION.

The ashes of aged stars murmur a narrative of cosmic recycling against the background of the Milky Way's weaving world, a tale in which stardust transforms into stellar nurseries. This chapter digs into the surprising legacy of the "old smokers," those mysterious red giants whose puffs of smoke are more than just a cosmic oddity; they are seeds for future generations of stars and planets.

Imagine a cosmic sculptor chiseling away at the fragments of old stars, shaping them into the foundations of future creation.

This is the job of the "old smokers," whose smoke is not a sign of their death, but rather an essential addition to interstellar dust, the raw material from which new stars and planets are formed.

The dust ejected into space by these celestial giants is more than just cosmic filth; it contains a rich mine of heavy elements formed in the scorching furnaces of previous generations of stars.

Elements like iron, silicon, and calcium, which are denser than their lighter counterparts, condense and cluster together, filling the interstellar medium with the necessary components for star formation.

This "legacy of heavy elements" idea depicts "old smokers" as cosmic couriers who transmit their richer composition farther out into the spiral arms of the Milky Way. But their contribution extends beyond mere richness.

The dust's unique makeup, which is heavy with "old smokers" debris, may have a significant impact on the birth of future stars. The temperature, density, and final size of the newborn stars that will emerge inside interstellar clouds are determined by the proper mix of heavy elements.

Stars formed from clouds enriched by "old smokers" may be more massive, have longer lifespans, and include more favorable circumstances for planetary formation.

The dance of dust influences not just star birth, but also planetary system formation. Heavy elements, such as those emitted by "old smokers," serve as cosmic glue, clumping together to create disks surrounding nascent stars. Planets are formed in these spinning disks, which resemble tiny galaxies.

However, the narrative of "old smokers" and their cosmic dust does not conclude with planetary creation. Their contribution affects the potential of existence itself.

Heavy elements like carbon, nitrogen, and oxygen, which are required for life compounds, are often found in abundance in dust enriched by these galactic oddballs.

The seeds of life, strewn across the cosmos, might very possibly have found fruitful ground in planetary systems formed by "old smoker" dust clouds. Understanding the influence of "old smokers" on interstellar dust and star formation is more than simply astronomical accounting; it holds the key to understanding the processes that led to the birth of our solar system and the possibility of life elsewhere in the cosmos.

By examining the smoke signals of these mysterious stars, we may trace the origins of the components that comprise our planet, our bodies, and the very air we breathe.

Further investigations at infrared and other wavelengths will reveal the rich world created by "old smokers" dust.

Analyzing the makeup of dust clouds, monitoring star-birth rates in areas enriched by their contributions, and looking for biosignatures on rocky planets produced from their leftovers are just a few of the interesting research avenues that lie ahead.

## CHAPTER 7: HOW THIS DISCOVERY CHALLENGES OUR UNDERSTANDING OF RED GIANT EVOLUTION AND STELLAR DEATH

Astronomers have seen the stately dance of cosmic evolution for millennia, observing stars bloom into vivid youth, simmer through middle age, and eventually fade into the twilight of red giants. This chapter, however, throws a celestial wrench in the works, investigating how the mysterious "old smokers" defy our accepted knowledge of red giant development and star death, compelling us to rewrite the cosmic recipe on the spot.

For red giants, the chapter was well-worn, describing their progressive depletion of fuel, enlargement into reddish circles, and gentle ejection of gas in their dying moments. The "old smokers," with their spectacular puffs of smoke and surprising behavior, are pulling out pages and scribbling in the margins, causing us to reconsider the whole red gigantic story.

One major problem is the "smoke" itself. The classic concept of red giants depicts them as gentle puffers that gradually lose bulk due to a consistent outflow of material. The "old smokers," on the other hand, are cosmic pranksters who throw theatrical tantrums and eject thick clouds of dust and gas in irregular spurts. This behavior points to a significantly more chaotic and dynamic interior terrain, a secret recipe far more intricate than our textbooks predicted.

Another wrinkle is the composition of the expelled material. Red giants were thought to lose their lighter components first, leaving a helium-rich shell. However, the "old smokers" expel heavy components with gusto, suggesting a deeper source of fuel or a distinct mixing mechanism inside their cores. This challenges our knowledge of stellar structure and the delicate balance of nuclear reactions in aged stars.

The "old smokers" also disrupt the timing of stellar death. Red giants were noted for their gradual, elegant demise, fading into obscurity over millions of years. However, these cosmic anomalies seem to have a flair for the theatrical, with smoke signals indicating occasional spurts of activity mixed with periods of relative stillness.

This calls into question our knowledge of the latter phases of star development, implying that hidden forces cause these outbursts and rewrite the death song of red giants.

However, the "old smokers" are not only tearing down but also building up. Their chaotic nature may hold the key to understanding how planets develop around red giants.

The expelled material, which is loaded with heavy elements, may supply the necessary ingredients for planets to consolidate and form, a prospect that was previously thought implausible near these dying stars.

This offers up fascinating new paths for investigating planetary formation and the possibility of life on red giants, which were previously assumed to be desolate wastelands.

Furthermore, the "old smokers" serve as cosmic cooks, enhancing the intergalactic soup with their heavy element-laden dust. This enriched material serves as a nursery for future generations of stars, impacting their composition, size, and capacity for life.

Understanding the effect of "old smoker" dust on star formation adds another element to the cosmic recipe, reminding us that the death of one generation serves as the cradle for the next.

Unraveling the secrets of the "old smokers" necessitates a new cookbook, one loaded with fresh ingredients and unique dishes. Astronomers are using advanced telescopes and cutting-edge simulations to gaze into the hearts of these mysterious stars, investigate the makeup of their dust clouds, and monitor their eruptions over time. Each new observation, each puff of smoke, adds another line to the updated formula, rewriting the tale of red giants and stellar death with every data point.

# CHAPTER 8: COULD "OLD SMOKERS" BE FOUND ELSEWHERE IN THE COSMOS?

The Milky Way is a thriving city of stars, each with its part in the vast cosmic drama. However, among the familiar faces, the "old smokers" stand out as intriguing crimson giants whispering secrets about the universe's hidden processes. This chapter goes beyond our galactic sanctuary, posing a fascinating question: may these celestial oddballs be found elsewhere in the universe, and if so, what would their existence reveal about the variety and development of galaxies?

Consider the cosmos to be a massive world made up of many hues and textures. Each galaxy, a unique patch on this fabric, represents the interaction of cosmic forces and the elements available at the time of its formation. The Milky Way, with its concentration of heavy elements and dense center area, may house unique celestial phenomena such as "old smokers," but are their murmurs limited to our galaxy or do they reverberate across the universe?

Unraveling this riddle requires us to go beyond our usual surroundings. Astronomers, equipped with large telescopes and complex simulations, are now scanning several galaxies for telltale evidence of "old smokers" among the cosmic crowd. One crucial characteristic is seen in the infrared spectrum.

The dusty puffs emitted by these mysterious stars shine brilliantly in this wavelength, perhaps revealing their existence even in faraway galaxies.

Finding "old smokers" in other galaxies would be more than simply a cosmic treasure hunt; it would have far-reaching consequences for our knowledge of galactic diversity. The existence of these stars in varied settings, with differing heavy element abundances and galaxy morphologies, may give information on the processes influencing their development and behavior.

Consider comparing recipes from other civilizations; similar meals might indicate common components and techniques, whilst dissimilar variants could highlight distinctive local influences.

Galaxies richer in heavy elements, such as the "metal-rich kitchen" of the galactic core, may be more likely to host "old smokers." This might disclose the history of star formation in that galaxy, including if it has experienced violent bursts of star birth in the past, filling the interstellar medium with the critical cosmic ingredients required for these mysterious stars to emerge.

Finding "old smokers" in galaxies significantly different from our own, maybe with calmer star formation histories or lower metal abundances, may call into question our present understanding of their creation processes.

It may drive us to rethink the cosmic recipe, looking for other ingredients or cooking techniques that may explain their presence in unexpected places.

The consequences go beyond stellar evolution. The dust emitted by "old smokers" serves as a galaxy fertilizer, replenishing the interstellar medium with heavy metals. This cosmic wealth then serves as raw material for future generations of stars and planets, perhaps influencing the circumstances for life to arise. Finding "old smokers" in other galaxies and examining their influence on interstellar dust composition might therefore reveal the possibility of life in other galactic settings, widening our quest for alien biosignatures.

The search for "old smokers" across galaxies is about more than just celestial bragging rights; it's about comprehending the universe's variety, the interaction of cosmic forces, and the possibility of life outside our galaxy. Each celestial whisper, each puff of smoke found in a faraway galaxy, contributes to the universal recipe, rewriting our concept of star formation, galactic evolution, and the possibility of life in the vast cosmic ocean.

## CHAPTER 9: THE IMPACT OF "OLD SMOKERS" ON THE CHEMICAL COMPOSITION AND HISTORY OF OUR GALAXY

A heavenly waltz takes place deep beneath the Milky Way's churning heart. Instead of elegant ballerinas, think dust and light, stars and smoke. At the heart of this cosmic dance are the mysterious "old smokers," red giants whose dramatic puffs defy our knowledge of star death and reveal mysteries about our galaxy's fundamental origins.

This chapter investigates the echoes these mysterious stars leave in the Milky Way's chemical makeup, tracking their influence on the cosmic soup from which future generations of stars and planets will emerge.

Consider the Milky Way to be a massive cosmic cauldron brimming with a variety of ingredients. Some, like hydrogen and helium, are cosmic stalwarts, playing a central role in the celestial drama. Others, such as the heavy metals created in the hot crucibles of previous generations of stars, provide a distinct taste to the mix. The "old smoker" myth revolves around these heavy elements, such as iron, silicon, and calcium, which serve as both dancers and stage in the cosmic waltz.

The "old smokers," formed in the galactic center's metal-rich furnace, serve as cosmic couriers, delivering their enriched composition out into the spiral arms. Their spectacular clouds of smoke, packed with heavy elements, serve as cosmic fertilizer, enriching the interstellar medium with essential components for future generations of stars and planets. According to this "legacy of heavy elements" notion, "old smokers" perform critical roles in galactic recycling, enabling the continuing circulation and enrichment of the Milky Way's basic building blocks.

However, the influence of "old smokers" extends beyond simple enrichment. The exact composition of the material they spew may have a significant impact on the properties of future stars. Imagine a chef meticulously picking spices for a delicate meal. The exact ratio of heavy elements controls the temperature, density, and even the final size of the fledgling stars that will develop inside the enriched clouds.

Stars formed from clouds containing "old smoker" material may be more massive, have longer lifespans, and have the ability to support more complicated planetary systems.

The dance of heavy elements influences not just star birth, but also the creation of planets. Picture a heavenly baker sifting flour and making dough. The heavy metals operate as binding agents, causing the spinning disks surrounding young stars to coalesce and ultimately form the rocky cores of future planets. Without the cosmic glue supplied by "old smokers," planetary systems may be smaller, less plentiful, and perhaps less favorable to life.

However, the echoes of "old smokers" extend beyond planetary formation. Their dusty puffs, which are high in carbon, nitrogen, and oxygen - the building blocks of living molecules - might represent the very seeds of life dispersed over the cosmic world. By enriching the interstellar medium with these critical components, "old smokers" may be unintentionally preparing the path for life to evolve on planets bathed in the light of future stars.

Understanding the influence of "old smokers" on the chemical composition of the Milky Way is more than just a cosmic accountancy exercise; it contains the key to understanding the events that led to the birth of our solar system, as well as the possibility of life elsewhere in space.

By researching the dust clouds speckled with their legacy, assessing the composition of stars created from their enriched environs, and looking for biosignatures on rocky planets formed from their leftovers, we may be able to see a reflection of our cosmic birth.

This chapter seeks to demonstrate the interconnectivity of the cosmos, as the ashes of old stars become the seeds for future creation. The "old smokers," enveloped in mysterious smoke, are more than just celestial anomalies; they play an important role in the Milky Way's epic tale. Their dramatic puffs resonate through the spiral arms, whispering tales about cosmic recycling, galactic development, and the possibility of life in the dust lanes.

So, let us keep listening to the echoes of the "old smokers," deciphering the mysteries of their cosmic smoke, and marveling at the universe's interconnection, where the past, present, and future are woven together in an elaborate world of celestial dust and cosmic whispers. As we probe further into their mysteries, we may be able to rewrite our galaxy's past, grasp the origins of our existence, and take a daring step closer to resolving the age-old question: are we alone in the universe?

The echoes of the "old smokers" summon us on a big cosmic adventure, not just to investigate the Milky Way's chemical world, but also to comprehend our role within it.

Their legacy serves as a reminder that even in the darkest of times, the whispers of the past hold the key to unlocking the mysteries of the future, leading us on an exciting voyage of discovery toward a better knowledge of our cosmic home and the possibility of life beyond.

# CHAPTER 10: THE LEGACY OF THE "OLD SMOKERS" AND THEIR POTENTIAL FOR FUTURE DISCOVERIES

The mysterious "old smokers" of the Milky Way, formerly celestial anomalies covered in cosmic dust, have revealed a world immensely more vibrant than we ever anticipated. Their cryptic puffs of smoke, previously a mystery, have now become a revelation, rewriting the scripts of star development and sparking a constellation of future discoveries.

This chapter honors the heritage of these mysterious stars while also exploring their exciting potential to reshape our knowledge of the cosmos.

The "old smokers" entered the stage with a stunning blast of smoke, defying our preconceived assumptions about red giants and stellar death.

Their chaotic behavior and disobedience of textbook assumptions sparked a scientific revolution, prompting us to rewrite the cosmic recipe and reconsider our knowledge of celestial dance. The "old smokers" left a significant legacy in the form of interstellar dust.

This dust is no longer just cosmic filth; it is a treasure mine of heavy metals formed in the flaming hearts of previous generations of stars. The "old smokers," acting as celestial couriers, carry this cosmic riches, enriching the interstellar medium and influencing the formation of future stars and planets. Understanding their involvement in this cosmic recycling scheme allows us to trace the origins of the components that comprise our planet, our bodies, and even the air we breathe.

However, the influence of "old smokers" extends beyond increasing the soup supply of star formation. Their dusty puffs may contain the key to understanding the creation of planetary systems orbiting red giants, which was previously thought implausible.

The heavy atoms they release serve as a glue, causing the spinning disks around young stars to cluster together and ultimately form the rocky cores of future planets. This offers fascinating new options for hunting for possibly habitable planets near these old celestial bodies, broadening the scope of our astrobiological search.

The "old smokers" also operate as cosmic storytellers, telling stories of star development on a grand scale. We may learn about the history of star formation in the Milky Way by examining the makeup of their dust clouds and the features of stars produced in their enriched surroundings.

Did the galaxy have bursts of strong star birth in the past, filling the area with heavy elements and preparing the way for the "old smokers" to emerge? Their drama serves as a window into the past, letting us put together the story of our cosmic neighbors. The disclosures made by the "old smokers" go beyond our galaxy. The quest for these mysterious stars in other, more complex galaxies might reveal a world of cosmic variety unlike anything we've ever seen.

Galaxies with diverse compositions, histories, and celestial dances may have their versions of "old smokers," who tell unique tales about their cosmic development and the possibility of life in their areas.

The search for these celestial curiosities throughout the universe transforms into a cosmic exploration voyage, an opportunity to compare and contrast, and rewrite the universal blueprint for star formation and life's evolution. However, the narrative of the "old smokers" does not conclude with discoveries; it starts with possibilities. Their presence provides interesting opportunities for future investigation.

New telescopes with greater eyesight and wider wavelength sensitivity will reveal the precise features of their smoke signals, track the structure of dust clouds, monitor star formation in richer settings, and look for biosignatures on possibly habitable worlds.

Each puff of cloud, each whisper of dust, has the potential to be a game-changing discovery, rewriting astronomy and astrobiology textbooks once again.

The "old smokers" left a legacy that demonstrates the force of inquiry, the excitement of discovery, and the universe's interconnection.

They remind us that even the most worn-out cosmic chapters may be rewritten, that, inconsequential quirks can reveal deep truths, and that the whispers of the unknown contain the key to unlocking the great mysteries of the universe.

Each discovery brings us closer to comprehending the universe's great tale, our position within it, and the possibility of life beyond our known frontiers.

The "old smokers" have rewritten the screenplay, and it is now up to us to continue the tale, to keep asking questions, to push the frontiers of our knowledge, and to remember that the most astounding insights often begin with a small puff of smoke in the darkness.

# CHAPTER 11: THE POWER OF INFRARED ASTRONOMY AND ITS FUTURE IN UNRAVELING COSMIC MYSTERIES

The cosmos speaks in many languages, yet certain secrets are only revealed to the most sensitive ears. These murmurs, cloaked by dust and concealed from the frigid grip of visible light, resound in the infrared spectrum. This chapter provides a window into this invisible world, honoring the power of infrared astronomy and its critical role in understanding the cosmic secrets hidden in the celestial darkness.

Imagine the universe as a consonance, with visible light performing the colorful melodies of stars and galaxies.

However, under the surface, deeper harmonic hums and a bassline are echoing from the universe's dusty corners. This is the music of infrared, a wavelength that is invisible to human sight yet may disclose heavenly mysteries that optical sensors cannot.

One of infrared's biggest advantages is its ability to penetrate dust. While dust scatters and absorbs visible light, infrared radiation penetrates, showing star nurseries wrapped in cosmic filth, fragile wisps of nascent galaxies veiled in cosmic dawn, and looking into the hearts of red giants buried by dusty shrouds. The "old smokers," mysterious stars whose dramatic puffs challenged our concept of stellar death, were only fully exposed via the delicate sight of infrared telescopes, demonstrating that sometimes the most profound discoveries are beyond the grasp of the naked eye.

Another significant advantage of infrared is its ability to detect heat. While visible light bounces off the surfaces of things, infrared displays their thermal glow, which is a clue of their interior activities. This enables scientists to investigate the secret furnaces of freshly formed stars, track the formation of planetary systems by examining the heat emitted by their dusty disks, and even detect the weak warmth of possibly habitable exoplanets circling distant suns.

The allure of infrared astronomy, however, extends beyond mere observation.

It functions as a cosmic interpreter, understanding the murmurs of dust. Astronomers can unlock the secrets of their composition by analyzing the spectrum of infrared light emitted by various celestial objects, revealing the presence of water ice in interstellar clouds, identifying the building blocks of planets within protoplanetary disks, and even looking for biosignatures like methane in exoplanet atmospheres.

Looking forward, infrared astronomy promises to take us into unexplored territory in the universe. Next-generation telescopes, such as the James Webb Space Telescope, will look deeper into the universe's early years, observing the first galaxies to ignite in the cosmic darkness and illuminating the cradles of stars and planets where life may take root.

These ground-breaking tools will also enable us to investigate the secrets of black holes, the cosmic maelstroms concealed beneath event horizons. By analyzing the heated material whirling around them, we may learn more about their feeding patterns, the influence they have on nearby galaxies, and perhaps the mysteries of gravity itself.

Furthermore, infrared technology is not limited to the skies. The same concepts are driving advancements in other scientific domains, such as medical imaging, which shows the inner workings of the human body, and environmental monitoring, which follows the Earth's changing climate.

Lessons from the cosmos are making their way back to our planet, improving our knowledge of the world around us and laying the way for a future in which technology inspired by the stars helps us confront some of humanity's most pressing concerns.

The voyage with infrared eyes is more than simply a scientific undertaking; it exemplifies the human spirit of exploration and discovery. It reminds us that the cosmos has mysteries beyond our immediate vision, that marvels lurk in the darkness, and that with the correct tools and a passion for knowledge, we may peel back the cosmic curtain and discover the hidden melody of the stars.

Let us keep pushing the frontiers of infrared technology, to improve our equipment, and to get a better grasp of the universe's infrared language. With each new whisper decoded, each cosmic secret revealed, we begin on a vast adventure of discovery, powered by the light that whispers in the darkness and motivated by the insatiable human yearning to comprehend the universe, ourselves, and our position within its majestic embrace.

# CHAPTER 12: UNDERSTANDING THE INTERCONNECTEDNESS OF STELLAR PROCESSES AND THEIR INFLUENCE ON THE UNIVERSE

The wide canvas of the night sky, with its glittering world of stars, may be seen as a collection of separate celestial bodies, each blazing brilliantly but alone. However, this chapter dives further, exposing the secret harmony behind this magnificent sight. It depicts the cosmos as a great orchestra, with stars playing unique notes that harmonize in a cosmic consonance, the echoes of which create the fabric of reality.

Consider a conductor who leads stars and supernovae rather than violins and cellos. This cosmic maestro directs the interior activities of every celestial body, determining its temperature, brightness, and longevity. Individual melodies are formed by the interaction of nuclear fusion, gravitational forces, and radiative energy inside stars, giving each celestial singer their character.

The powerful red giant plays an important role in this cosmic consonance. As these elderly stars approach the conclusion of their brilliant hydrogen-burning era, they inflate into reddish spheres, and their interior processes change radically. Helium fusion occurs in their cores, while their outer layers expand and cool, ejecting gas and dust in magnificent puffs that enrich the interstellar environment.

These "old smokers," as they've been dubbed, are critical components of the cosmic soup from which future generations of stars and planets will emerge.

But the consonance goes beyond these particular notes. Red giants spew dust and gas-rich with heavy elements formed in their hot hearts, which serve as raw ingredients for future cosmic compositions. According to this "legacy of heavy elements" idea, red giants play an important role in galactic recycling, enriching the interstellar medium and affecting the nature of future stars and planets. Stars formed from clouds containing "old smoker" material may be more massive, have longer lifetimes, and have the potential for more complex planetary systems.

The consonance of stellar activities also plays an important part in the development of life. Carbon, nitrogen, and oxygen, which are necessary for life compounds, are often found in abundance in dust enriched by red giants. The seeds of life, which are strewn across the cosmos, may have found fruitful ground in planetary systems formed by cosmic dust storms. Understanding the effects of star activities on interstellar dust composition is therefore an important step in the search for alien life.

Even the tragic deaths of stars add to the cosmic consonance. Supernovae, the catastrophic end of big stars, release a flood of energy and heavy materials, enriching the interstellar medium on a gigantic scale.

These cosmic cataclysms cause star formation in neighboring clouds, replenishing the galaxy with new celestial actors and influencing the distribution of elements required for future planetary systems. Supernovae are more than simply devastating finales; they are explosive transitions required for the continuing of the cosmic consonance.

However, the consonance is not limited to individual stars; it extends to the vast size of galaxies. The pace of star formation, the makeup of the interstellar medium, and the existence of heavy metals change amongst galaxies, resulting in distinctive celestial choirs with diverse timbres. The study of galactic-scale differences helps us to get a better understanding of the universe's variety, the many routes of star development, and the possibility of life in various cosmic conditions.

The consonance of the stars not only plays in the past and present but also influences the future. Understanding the interdependence of stellar processes allows us to anticipate the future development of galaxies, the possibility of star formation in certain locations, and even the chance of habitable planets forming.

This understanding allows us to hunt for exoplanets more effectively, comprehend the circumstances required for life, and perhaps answer the age-old question: are we alone in the universe?

The universe's consonance, with its magnificent melodies and beautiful interaction, demonstrates the interconnection of everything. The solitary stars are closely connected, with their activities impacting one another over enormous distances, altering galaxies' history and seeding the possibility for life. As scientists continue to comprehend this cosmic song, we learn more about the secrets of star development and galaxy variety, as well as our role in the universe's big orchestra.

Therefore, let us pay attention to the stars, to red giant whispers, supernova explosions, and galaxies' beautiful melodies. Let their light lead us through the dark, their processes enlighten the unknown, and their consonance motivates us to continue our search for cosmic understanding. By unraveling the melody of the stars, we may just discover the answer to the most fundamental question of all: the vast consonance of existence, in which every note, every celestial dance, plays an important part in the developing song of the universe.

# CHAPTER 13: THE HUMAN QUEST TO UNDERSTAND OUR PLACE IN THE COSMOS THROUGH STELLAR DISCOVERIES

Gazing up at the star-spangled canvas, a primordial question arises: where do we fit into this huge, glittering world? From ancient astronomers tracing constellations to current telescopes looking into the cosmic cradle, humanity's quest for our position in the universe is intertwined with celestial discoveries.

This chapter digs into this long-standing pursuit, examining how understanding the lives and deaths of stars offers insight into our beginnings and the possibility of cosmic kinship.

Consider the cosmos to be a great family gathering, with enormous celestial halls filled with stars, galaxies, and the rich world of life that sprang from their flaming hearts. Our hunt for cosmic kinfolk is analogous to looking for cousins, aunts, and uncles, tracing the genealogy of the elements that comprise our bodies back to the nucleosynthesis furnaces of distant stars. Each celestial discovery speaks a piece of our family history, revealing links stronger than blood and forged in the furnace of creation itself.

One important episode in this cosmic story focuses on the "old smokers," mysterious red giants whose spectacular plumes of dust defy our knowledge of star death.

Heavy metals ejected from their hot cores serve as the foundation for future generations of stars and planets. Our entire existence, the silicon in our bones, the iron in our blood, might be a result of these celestial anomalies, providing a concrete connection to our cosmic ancestors.

However, the affinity extends beyond the common elements. Understanding the various lifespans and activities of stars helps us understand the circumstances required for life to arise. Massive stars, with their cataclysmic supernovae, initiate the cosmic enrichment process, but smaller stars, such as the Sun, provide long-term stability, which is essential for the emergence of sophisticated life on orbiting planets.

Studying the many ways stars live and die helps us comprehend the cosmic ecosystem, the delicate balance of forces that may enable life to spread to other parts of the cosmos.

The hunt for alien life itself feeds our desire for cosmic brotherhood. Planets circling exoplanets, bathed in the light of faraway stars, have the enticing prospect of sheltering our cosmic brothers.

Understanding the influence of various star types on planetary environments helps us focus our hunt for habitable zones or sweet spots where liquid water, the elixir of life, may exist. Each verified exoplanet, or rocky world circling a Sun-like star, becomes a prospective cousin, a ray of hope in the vast cosmic ocean.

However, our cosmic brotherhood goes beyond biological existence. The interaction of gravity, electromagnetic, and the basic forces that regulate stars also affects the movement of atoms inside us. The exact mechanisms that govern stellar development, from hydrogen fusion to supernovae, occur on a microscopic scale inside our bodies, reminding us that we are microcosms of the macrocosm, little stars dancing to the same heavenly music.

The search for cosmic kinship is about more than simply discovering extraterrestrial life; it is about seeing ourselves reflected in the vast cosmic mirror.

Understanding the lives and deaths of stars gives us a greater respect for the fragility and value of our own life on Earth. The transitory nature of solar flares teaches us to appreciate our Sun's stability, yet the immensity of the universe humbles us, reminding us of our position in the great fabric of existence.

This quest for kindred isn't a solo one; it's a worldwide consonance of inquiry and cooperation. Astronomers from all backgrounds collaborate to solve the universe's riddles, using powerful telescopes and cutting-edge technology. Each discovery, each decoded star whisper, is a cause for joy, bringing us one step closer to comprehending our position in the universe.

So, let us continue on this epic adventure, driven by the need for connection and the joy of discovery. Let the constellations serve as our celestial maps, the stars as guiding lights, and the echoes of supernovae as rallying calls.

As we investigate the lives and deaths of stars, we may discover our relatives among the exoplanets, our kin in the dance of atoms, and, finally, our position as a child of the universe, permanently tied to the consonance of the stars.

# CONCLUSION

The "old smokers" have vanished from the stage, their spectacular puffs of dust drifting into the cosmic vacuum. However, their echoes reverberate long after they've vanished, revealing mysteries about the cosmos and sparking a constellation of questions that will guide our cosmic journey for millennia to come. This last chapter is not a farewell, but a meditation on the remaining embers of their legacy, a reminder that the most profound discoveries can raise more questions than answers.

Consider the "old smokers" not as celestial anomalies, but as cosmic poets, their dusty poems written over the galactic canvas.

Each puff, each wisp of enriched matter, is a line in a magnificent poem about star development retold in the language of dust and light. We, the cosmic readers, are still decoding these lyrical stanzas, gaining insights about the dance of life and death inside red giants, their influence on galactic development, and the possibility of life beyond our solar system.

One notable echo of the "old smokers" is their challenge to our knowledge of stellar death. It's no longer just a peaceful slip into oblivion. Their spectacular outbursts and defiance of textbooks foreshadow a vibrant and chaotic last act, a cosmic fireworks show powered by secret processes inside their aged hearts.

Understanding these processes and decoding the secrets concealed inside their hazy whispers has the potential to rewrite the whole story of stellar death, presenting a much more complicated and diversified consonance of heavenly ends.

The "old smokers" leave us with a fresh understanding of the universe's interconnection. Their dusty puffs, laden with heavy metals, serve as cosmic messengers, enriching the interstellar medium and influencing the formation of future stars and planets. We are all actually stardust, formed from the ashes of past suns and linked by an unseen network woven from the remains of stellar explosions.

This cosmic connection, created in the blazing furnaces of red giants, reminds us that our existence is a reflection of those celestial anomalies, demonstrating the interdependence of all things.

However, the murmurs of the "old smokers" transcend beyond scientific insights, instilling a feeling of curiosity, a need for inquiry, and a desire for connection. Their sheer presence motivates us to improve our equipment, broaden our knowledge of the universe's secret processes, and pursue alien life with fresh zeal. Each newly found exoplanet circling a Sun-like star, or each puff of dust identified near a distant red giant provides a fascinating clue in the cosmic detective narrative of life's origin.

So, when the embers of the "old smokers" fade, the journey they started continues. Our telescopes scan the heavens with hungry eyes, looking for echoes of their drama in other cosmic corners, hoping to see cousins, aunts, and uncles at the celestial family reunion.

The questions kids ask are our guiding lights: what motivates their outbursts? How do they affect the creation of planetary systems? Can life thrive in the shadow of such tumultuous stars?

These issues are more than just intellectual exercises; they are threads sewn into the fabric of our own lives. By deciphering the secrets of the "old smokers," we acquire a better grasp of the world we live in, our position within it, and the cosmic forces that affect our very existence. In their murmurs, we hear echoes of our beginnings, the narrative of stardust coming to life, and the enticing prospect of cosmic brotherhood among the stars.